出版説明 _____

　　本館一向倡導優質閱讀,近年來連續推出了以"Q"為標識的 "Quality English Learning 優質英語學習"系列,其中《讀名著學英語》叢書,更是香港書展入選好書,讀者反響令人鼓舞。推動社會閱讀風氣,推動英語經典閱讀,藉閱讀拓廣世界視野,提高英語水平,已經成為一種潮流。

　　然良好閱讀習慣的養成非一日之功,大多數初、中級程度的讀者,常視直接閱讀厚重的原著為畏途。如何給年輕的讀者提供切實的指引和幫助,如何既提供優質的學習素材,又提供名師的教學方法,是當下社會關注的重要問題。針對這種情況,本館特別延請香港名校名師,根據多年豐富的教學經驗,精選海外適合初、中級英語程度讀者的優質經典讀物,有系統地出版了這套叢書,名為《Black Cat 優質英語階梯閱讀》。

　　《Black Cat 優質英語階梯閱讀》體現了香港名校名師堅持經典學習的教學理念,以及多年行之有效的學習方法。既有經過改寫和縮寫的經典名著,又有富創意的現代作品;既有精心設計的聽、説、讀、寫綜合練習,又有豐富的歷史文化知識;既有彩色插圖、繪圖和照片,又有英美專業演員朗讀作品的 CD。適合口味不同的讀者享受閱讀之樂,欣賞經典之美。

　　《Black Cat 優質英語階梯閱讀》由淺入深,逐階提升,好像參與一個尋寶遊戲,入門並不難,但要真正尋得寶藏,需要投入,更需要堅持。只有置身其中的人,才能體味純正英語的魅力,領略得到真寶的快樂。當英語閱讀成為自己生活的一部分,英語水平的提高自然水到渠成。

商務印書館(香港)有限公司
編輯部

使用説明

❶ 應該怎樣選書？

按閱讀興趣選書

《Black Cat 優質英語階梯閱讀》精選世界經典作品，也包括富於創意的現代作品；既有膾炙人口的小説、戲劇，又有非小説類的文化知識讀物，品種豐富，內容多樣，適合口味不同的讀者挑選自己感興趣的書，享受閱讀的樂趣。

按英語程度選書

《Black Cat 優質英語階梯閱讀》現設 Level 1 至 Level 6，由淺入深，涵蓋初、中級英語程度。讀物分級採用了國際上通用的劃分標準，主要以詞彙（vocabulary）和結構（structures）劃分。

Level 1 至 Level 3 出現的詞彙較淺顯，相對深的核心詞彙均配上中文解釋，節省讀者查找詞典的時間，以專心理解正文內容。在註釋的幫助下，讀者若能流暢地閱讀正文內容，就不用擔心這一本書程度過深。

Level 1 至 Level 3 出現的動詞時態形式和句子結構比較簡單。動詞時態形式以現在時（present simple）、現在時進行式（present continuous）、過去時（past simple）為主，句子結構大部分是簡單句（simple sentences）。此外，還包括比較級和最高級（comparative and superlative forms）、可數和不可數名詞（countable and uncountable nouns）以及冠詞（articles）等語法知識點。

Level 4 至 Level 6 出現的動詞時態形式，以現在完成時（present perfect）、現在完成時進行式（present perfect continuous）、過去完成時（past perfect continuous）為主，句子結構大部分是複合句（compound sentences）、條件從句（1st and 2nd conditional sentences）等。此外，還包括情態動詞（modal verbs）、被動形式（passive forms）、動名詞（gerunds）、

Washington Irving

Rip Van Winkle and The Legend of Sleepy Hollow

睡谷傳奇

商務印書館

This Chinese edition of *Rip Van Winkle and The Legend of Sleepy Hollow* has been published with the written permission of Black Cat Publishing.

The copyright of this Chinese edition is owned by The Commercial Press (H.K.) Ltd.

Name of Book: Rip Van Winkle and The Legend of Sleepy Hollow
Author: Washington Irving
Retold by: Cheryl Jefferson
Editors: Claudia Fiocco, Emma Berridge
Design: Nadia Maestri
Illustrations: Alfredo Belli
Edition: © 2003 Black Cat Publishing
an imprint of Cideb Editrice, Genoa, Canterbury

系 列 名：Black Cat 優質英語階梯閱讀 · Level 1
書　　名：睡谷傳奇
責任編輯：傅　伊
封面設計：張　毅
出　　版：商務印書館（香港）有限公司
　　　　　香港筲箕灣耀興道 3 號東滙廣場 8 樓
　　　　　http://www.commercialpress.com.hk
發　　行：香港聯合書刊物流有限公司
　　　　　香港新界大埔汀麗路 36 號中華商務印刷大廈 3 字樓
印　　刷：中華商務彩色印刷有限公司
　　　　　香港新界大埔汀麗路 36 號中華商務印刷大廈
版　　次：2015 年 1 月第 6 次印刷
　　　　　© 2004 商務印書館（香港）有限公司
　　　　　ISBN 978 962 07 1689 8
　　　　　Printed in Hong Kong

短語動詞（phrasal verbs）等語法知識點。

　　根據上述的語法範圍，讀者可按自己實際的英語水平，如詞彙量、語法知識、理解能力、閱讀能力等自主選擇，不再受制於學校年級劃分或學歷高低的約束，完全根據個人需要選擇合適的讀物。

② 怎樣提高閱讀效果？

　　閱讀的方法主要有兩種：一是泛讀，二是精讀。兩者各有功能，適當地結合使用，相輔相成，有事半功倍之效。

　　泛讀，指閱讀大量適合自己程度（可稍淺，但不能過深）、不同內容、風格、體裁的讀物，但求明白內容大意，不用花費太多時間鑽研細節，主要作用是多接觸英語，減輕對它的生疏感，鞏固以前所學過的英語，讓腦子在潛意識中吸收詞彙用法、語法結構等。

　　精讀，指小心認真地閱讀內容精彩、組織有條理、遣詞造句又正確的作品，着重點在於理解 "準確" 及 "深入"，欣賞其精彩獨到之處。精讀時，可充分利用書中精心設計的練習，學習掌握有用的英語詞彙和語法知識。精讀後，可再花十分鐘朗讀其中一小段有趣的文字，邊唸邊細心領會文字的結構和意思。

　　《Black Cat 優質英語階梯閱讀》中的作品均值得精讀，如時間有限，不妨嘗試每兩個星期泛讀一本，輔以每星期挑選書中一章精彩的文字精讀。要學好英語，持之以恆地泛讀和精讀英文是最有效的方法。

③ 本系列的練習與測試有何功能？

　　《Black Cat 優質英語階梯閱讀》特別注重練習的設計，為讀者考慮周到，切合實用需求，學習功能強。每章後均配有訓練聽、說、讀、寫四項技能的練習，分量、難度恰到好處。

聽力練習分兩類，一是重聽故事回答問題，二是聆聽主角對話、書信朗讀、或模擬記者訪問後寫出答案，旨在以生活化的練習形式逐步提高聽力。每本書均配有 CD 提供作品朗讀，朗讀者都是專業演員，英國作品由英國演員錄音，美國作品由美國演員錄音，務求增加聆聽的真實感和感染力。多聆聽英式和美式英語兩種發音，可讓讀者熟悉二者的差異，逐漸培養分辨英美發音的能力，提高聆聽理解的準確度。此外，模仿錄音朗讀故事或模仿主人翁在戲劇中的對白，都是訓練口語能力的好方法。

閱讀理解練習形式多樣化，有縱橫字謎、配對、填空、字句重組等等，注重訓練讀者的理解、推敲和聯想等多種閱讀技能。

寫作練習尤具新意，教讀者使用網式圖示（spidergrams）記錄重點，採用問答、書信、電報、記者採訪等多樣化形式，鼓勵讀者動手寫作。

書後更設有升級測試（Exit Test）及答案，供讀者檢查學習效果。充分利用書中的練習和測試，可全面提升聽、說、讀、寫四項技能。

④ 本系列還能提供甚麼幫助？

《Black Cat 優質英語階梯閱讀》提倡豐富多元的現代閱讀，巧用書中提供的資訊，有助於提升英語理解力，擴闊視野。

每本書都設有專章介紹相關的歷史文化知識，經典名著更有作者生平、社會背景等資訊。書內富有表現力的彩色插圖、繪圖和照片，使閱讀充滿趣味，部分加上如何解讀古典名畫的指導，增長見識。有的書還提供一些與主題相關的網址，比如關於不同國家的節慶源流的網址，讓讀者多利用網上資源增進知識。

Contents

Rip Van Winkle
瑞普・凡・溫克爾

The Legend of Sleepy Hollow
睡谷傳奇

🎧 This symbol indicates a listening activity.
聽力練習的標記

ABOUT THE AUTHOR

Name: Washington Irving,
"The Father of American Literature"
Born: In New York City
When: April 3, 1783
First Book:
Knickerbocker's History of New York (1809)
Most famous stories:
Rip Van Winkle and
The Legend of Sleepy Hollow (1819)
Travels: England, Italy,
Spain, France
Died: 1859

American Painters and the Catskill Mountains

The Clove, Catskills (ca. 1827) by Thomas Cole (1801-1848), New Britain Museum of American Art, Connecticut.

The story of *Rip Van Winkle* takes place in the Catskill Mountains, New York.

These splendid mountains inspire [1] many American painters.

What colors do you see in this painting? ....

....

What season is it? ....

....

What's the weather like? ....

Circle the adjective(s) that best describe(s) the painting:

> ugly   mysterious [2]   serene [3]   beautiful   colorful

1. **inspire** : 給⋯以靈感。
2. **mysterious** : 神秘的。
3. **serene** : 寧靜的。

*An old inn from the 1700s,
similar to [1] the one in Rip Van Winkle's village.*

*A one-room school like the one where
Ichabod Crane teaches.*

1. **similar to**：與⋯相似的。

THE THIRTEEN ORIGINAL COLONIES, 1776.

MAINE

NEW HAMPSHIRE

MASSACHUSETTS

NEW YORK

RHODE ISLAND

CONNECTICUT

PENNSYLVANIA

NEW JERSEY

DELAWARE

VIRGINIA

NORTH CAROLINA

SOUTH CAROLINA

GEORGIA

Atlantic Ocean

N
W E
S

Rip Van Winkle

1 Here are some words from the story. Do you know them?

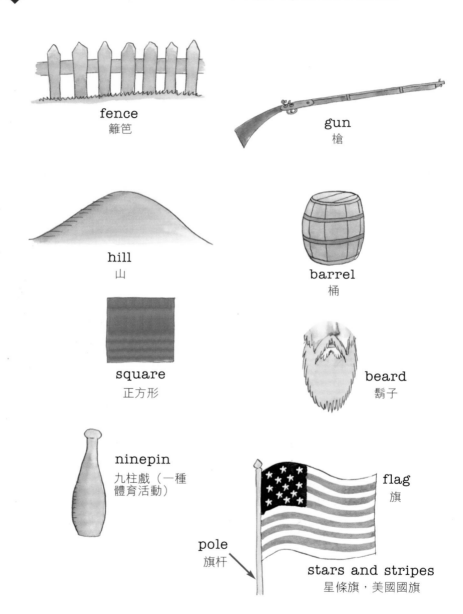

fence
籬笆

gun
槍

hill
山

barrel
桶

square
正方形

beard
鬍子

ninepin
九柱戲（一種
體育活動）

flag
旗

pole
旗杆

stars and stripes
星條旗，美國國旗

14

CHAPTER ONE

The Van Winkle Family

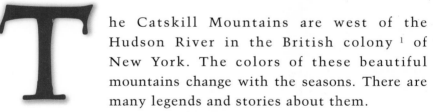

The Catskill Mountains are west of the Hudson River in the British colony [1] of New York. The colors of these beautiful mountains change with the seasons. There are many legends and stories about them.

The year is 1765. Rip Van Winkle lives in a village at the foot of the Catskill Mountains. The village is of Dutch [2] origin. [3]

Rip lives in an old, ugly house. He is a simple, kind man. Everyone in the village loves him because he is friendly and helpful. He plays with the children and makes things for them.

Rip's wife is very harsh [4] with him. She nags [5] him all the time.

---

1. **colony** : 殖民地。
2. **Dutch** : 荷蘭人的。
3. **origin** : 出身。
4. **harsh** : 苛刻的。
5. **nags** : 抱怨。

Rip Van Winkle

However, he is patient [1] and obedient. [2] He never gets angry at his wife. He simply keeps away [3] from her.

Rip has only one bad point: he doesn't like working. He only does the things he likes. He sits on a rock and fishes all day. He walks in the forest and hunts animals. He builds fences for his friends. He helps the women in their homes and gardens. He does everyone's work but not his own. His farm is the worst in the village. He never looks after his land, his garden or his house. His fences are broken. His cow eats the vegetables in his garden. His children's clothes are old and dirty.

Mrs Van Winkle nags him about his laziness. [4] Poor Rip never says anything. He has a dog called Wolf. Mrs Van Winkle doesn't like Wolf and Wolf doesn't like her. She often throws things at the poor animal.

Rip often goes to the village inn. [5] It is called "King George the Third" in honor of [6] the British king. There is a picture of King George outside the inn.

1. **patient**：忍耐的。
2. **obedient**：服從的。
3. **keeps away**：避開。
4. **laziness**：懶惰。
5. **inn**：酒館。
6. **in honor of**：紀念。

Rip Van Winkle

The men of the village sit outside the inn during the summer. They talk and drink. They listen carefully to the school teacher, Derrick Van Bummel. He is an educated [1] man and uses long words.

Nicholas Vedder is the innkeeper. [2] He sits outside the inn from morning until night.

Rip likes going to the inn and talking to his friends. He's happy there. But Mrs Van Winkle doesn't like the inn or his friends. "This is a place for lazy men!" she says. "You must never go there again, Rip!"

1. **educated**：受過良好教育的。
2. **innkeeper**：酒館老闆。

UNDERSTANDING THE TEXT

KET

 1 **Are these sentences "Right" (A) or "Wrong" (B)? If there is not enough information to answer, choose "Doesn't say" (C). Circle the correct answers.**

1. There are many legends about the Catskill Mountains.

    A Right   **B** Wrong   **C** Doesn't say

2. Rip Van Winkle lives in an ugly house and wears old clothes.

    A Right   **B** Wrong   **C** Doesn't say

3. No one in the village likes Rip.

    A Right   **B** Wrong   **C** Doesn't say

4. Mrs Van Winkle's name is Anna.

    A Right   **B** Wrong   **C** Doesn't say

5. Rip does everyone's work but not his own.

    A Right   **B** Wrong   **C** Doesn't say

6. Wolf likes Mrs Van Winkle.

    A Right   **B** Wrong   **C** Doesn't say

7. Derrick Van Bummel is a school teacher.

    A Right   **B** Wrong   **C** Doesn't say

8. The inn is called "King George the Third" in honor of the American king.

    A Right   **B** Wrong   **C** Doesn't say

9. It is a very big inn.

    A Right   **B** Wrong   **C** Doesn't say

10. The innkeeper says, "You must never come to the inn again, Rip."

    A Right   **B** Wrong   **C** Doesn't say

2 **Listen to the following conversation twice. Then tick (✓) the correct answer A, B or C.**

1. Rip's fences are
- **A** ☐ old.
- **B** ☐ broken.
- **C** ☐ ugly.

2. Rip's friends are
- **A** ☐ Nicholas and Derrick.
- **B** ☐ George and Nicholas.
- **C** ☐ Derrick and George.

3. Rip wants to build the fences on
- **A** ☐ Monday.
- **B** ☐ Thursday.
- **C** ☐ Tuesday.

4. Mrs Van Winkle says, "You must build the fences on
- **A** ☐ Thursday".
- **B** ☐ Monday".
- **C** ☐ Wednesday".

5. Mrs Van Winkle doesn't like Nicholas because he is
- **A** ☐ lazy.
- **B** ☐ old.
- **C** ☐ ugly.

6. On Wednesday Rip must
- **A** ☐ build the fences.
- **B** ☐ go to the market
- **C** ☐ work in the garden.

7. Rip must go to the market on
- **A** ☐ Monday.
- **B** ☐ Thursday.
- **C** ☐ Wednesday.

8. What time is the market?
- **A** ☐ 10:00 a.m.
- **B** ☐ 9:30 a.m.
- **C** ☐ 10:15 a.m.

A Strange Afternoon

O ne autumn day Rip goes up the Catskill Mountains. After walking for many hours he feels tired. He sits on a green hill and looks at the majestic [1] Hudson River and as he is looking at the sky he says, "It's almost evening. I must go home. My wife always gets angry when I'm late."

As he's going down the mountain he hears a voice, "Rip Van Winkle! Rip Van Winkle!"

He looks around but he sees no one. He hears someone call, "Rip Van Winkle! Rip Van Winkle!" again. Then he sees a very strange man. He is climbing up the hill with something on his back. Rip thinks, "Who is this? No one ever comes here. Perhaps I can help him."

---

1. **majestic** : 壯麗的。

21

Rip Van Winkle

Rip walks toward [1] the man and is very surprised. He sees a short, square [2] old man. He has long, white hair and a beard.

Rip thinks, "How bizarre! [3] He's wearing very old clothes. They are Dutch clothes of the 1600s! Who is this man?"

The old man is carrying a small barrel on his back. Rip helps the man and carries his barrel. Together they go up the valley. Rip can hear the thunder [4] but he can't see the storm.

Rip follows the strange man. He thinks, "Where are we going and why?" Suddenly, they walk into a small amphitheater. [5] It is made of big stones and trees. There are some strange men there.

---

1. **toward**：朝着。
2. **square**：健壯的。
3. **bizarre**：古怪。
4. **thunder**：雷聲。
5. **amphitheater**：圓形露天競技場。

A Strange Afternoon

They are all wearing Dutch clothes of the 1600s too and they all have beards. There is a captain, [1] he is wearing a captain's hat. He is a big, fat, round man.

They are playing ninepins. The balls sound the same as thunder. No one is speaking. The mysterious silence and the cold eyes of the men scare [2] Rip. The old man points to the barrel and to the other old men. "These men want to drink," Rip thinks. He serves them and then they continue playing again.

Rip drinks from the barrel too. It is very strong but Rip likes it. He drinks again and again. Soon he falls asleep.

---

1. **captain**：船長。
2. **scare**：使恐懼。

UNDERSTANDING THE TEXT

1 **Choose the correct word to complete each sentence. Tick (✓) A, B or C.**

1. Rip goes up the Catskill Mountains in the

    A ☐ spring.

    B ☐ autumn.

    C ☐ summer.

2. When he goes down the mountain he hears

    A ☐ thunder.

    B ☐ strange music.

    C ☐ a voice.

3. Then he sees

    A ☐ a short, square old man with a small barrel.

    B ☐ a tall young man with a small barrel.

    C ☐ the captain of a ship.

4. Rip follows the strange man to

    A ☐ the Catskill Mountains.

    B ☐ the Hudson River.

    C ☐ a small amphitheater.

5. What are the strange men wearing?

    A ☐ Dutch clothes of the 1600s.

    B ☐ Old, dirty clothes.

    C ☐ Black clothes.

6. Their game of ninepins

   A ☐ is fun.

   B ☐ makes the same noise as thunder.

   C ☐ is silent and mysterious.

7. Rip drinks from the barrel and soon he

   A ☐ returns home.

   B ☐ starts playing ninepins.

   C ☐ falls asleep.

2 **Match these opposites.**

1. short	**a.** early
2. square	**b.** young
3. small	**c.** tall
4. fat	**d.** round
5. noise	**e.** thin
6. late	**f.** big
7. old	**g.** silence

Look at page 22 and read the description of the strange man. Now write to your new pen pal and use some of the words above to describe yourself and your best friend. Start like this:

I am ....

....

....

My best friend ....

....

....

Fill in the gaps with the correct prepositions. Some prepositions can be used twice.

around   into   up   from   on   toward   of

a. One day Rip climbs .... the Catskill Mountains.

b. He sits .... a green hill.

c. He looks .... him but he sees no one.

d. The man has something .... his back.

e. Rip walks .... the man.

f. They walk .... a small amphitheater.

g. It is made .... stones and trees.

h. Rip drinks .... the barrel.

 The secret message

How many words can you find in the word river? Circle them with a red pencil and write the secret message below. This secret message will help you understand Chapter 3!

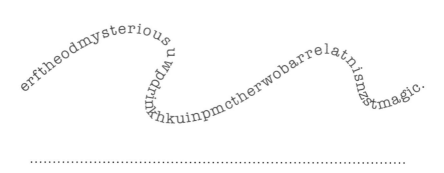

....

The Next Morning

Rip wakes up the next morning and says, "It's morning!" Then he remembers the old man with the barrel, the game of ninepins and the strong drink. "My wife will be furious.[1] What can I tell her?"

He looks for[2] his gun and finds an old gun in its place. He calls his dog, but the dog does not come.

"Those old men have my gun and my dog. I must find them."

He gets up and tries to walk. "It's difficult to walk. My poor legs!" he thinks.

He cannot find the old men or his dog. He is very hungry and decides to go home. He starts going down the mountain and when he is near the village he meets some people.

---

1. **furious** : 非常生氣的。
2. **looks for** : 尋找。

Rip Van Winkle

"How strange! I know everyone in the village, but I don't know these people. Their clothes are very different," thinks Rip.

The people look at him and his clothes. They are surprised and they touch their faces.

"Why are they touching their faces?" Rip thinks. He touches his face and is surprised. He has a very long beard!

When Rip arrives at the village the children point at his gray beard and the dogs bark [1] at him. The village looks different. It is very big. There are new houses with new names on the doors. He sees new faces everywhere.

Everything is changed. Poor Rip is very confused. [2]

"How is this possible? This is MY village. I can see the Catskill Mountains and the Hudson River. But everything is different. It must be the effect [3] of the strong drink," Rip thinks.

---

1. **bark**：（狗）吠，叫。
2. **confused**：困惑的。
3. **effect**：影響。

The Next Morning

He finds his house and goes in. It is abandoned [1] and all the windows are broken and there is no door. There is nothing in the house anymore. He calls for his wife and children. But they don't answer him.

"What is happening? Where is my family?" he thinks. He goes to the inn. He sees a big hotel in its place. Rip reads the name on the door:

He sees a tall pole in front of the hotel. On the pole there is a flag. There are stars and stripes on it. He goes inside the inn and looks for the picture of King George. He finds it but it is different. King George is wearing a blue coat instead of a red one and he is wearing a hat instead of a crown. Under the picture Rip sees the words "George Washington".

---

1. **abandoned** : 無人居住的。

UNDERSTANDING THE TEXT

 Read the paragraph and write the best word A, B or C in each space. The first one is done for you.

Rip wakes up ¹......B...... sunny morning. He looks ²..............
his gun and his dog.

He gets up and ³.............. to walk. He goes ⁴.............. the
mountain and ⁵.............. some people. The people look at
⁶.............. and ⁷.............. clothes. Rip arrives ⁸.............. the
village. Everything ⁹.............. different. He ¹⁰.............. his
house. There is ¹¹.............. in the house. The old inn
¹².............. a new name. ¹³.............. a pole there ¹⁴..............
a flag with stars and stripes.

1. A in	**B** one	**C** at
2. A to	**B** at	**C** for
3. A tries	**B** try	**C** trys
4. A in	**B** up	**C** down
5. A meet	**B** meets	**C** met
6. A his	**B** him	**C** he
7. A his	**B** he	**C** him
8. A at	**B** to	**C** on
9. A are	**B** be	**C** is
10. A to find	**B** find	**C** finds
11. A nothing	**B** anything	**C** none
12. A is	**B** has	**C** have
13. A On	**B** At	**C** In
14. A are	**B** be	**C** is

2 **Do you know the five senses? Match the numbers with the letters.**

1. You see with your...   **a.** ears
2. You taste with your...   **b.** nose
3. You hear with your...   **c.** fingers
4. You touch with your...   **d.** eyes
5. You smell with your...   **e.** tongue

3 **Help Rip find his house. Write down the names of the objects in the maze from the story. How many words can you find with the letter 'h' in them?**

31

CHAPTER FOUR

The Spirits of the Catskill Mountains

There are some people standing near the door of the Union Hotel. But Rip does not know anyone. They are not quiet, slow and sleepy. [1] They all seem noisy and fast. [2] They talk quickly in loud voices.

"Where are my old friends?" Rip thinks.

Rip sees a thin man with many papers in his pockets. [3] He is talking about political elections, [4] liberty, [5] the war of seventy-six and Congress. [6] These new words confuse Rip.

1. **sleepy** : 睏倦的，想睡的。
2. **fast** : 放蕩的。
3. **pockets** : 衣袋。
4. **elections** : 選舉。
5. **liberty** : 自由。
6. **Congress** : （美國）國會。

The Spirits of the Catskill Mountains

"What is this man saying? What is 'Congress'? Is it a town? What war of seventy-six?" Rip thinks.

Soon everyone is looking at Rip's long beard and strange clothes. The thin man asks him, "Are you a Federal [1] or a Democrat?" [2]

Rip does not understand these words. An old gentleman asks Rip, "Why do you have a gun? Do you want to fight?"

"Oh, no, no!" says Rip. "I am a poor, quiet man, and King George is my king."

"What! King George!" the people cry. "This man is a spy! He's not an American. Who is he? Call the police!"

The old gentleman asks, "Why are you here?"

"I'm looking for my friends. But I can't find them," Rip says.

"Who are your friends?" asks the old gentleman.

"Nicholas Vedder," says Rip.

"Nicholas Vedder? He's dead!"

"And Brom Dutcher and Derrick Bummel."

"Dutcher is dead and Bummel is a Congressman," [3] says the old gentleman.

Rip is unhappy and confused. "I am alone in a strange world and no one knows me. Does anyone know Rip Van Winkle?" he cries.

"Oh, Rip Van Winkle!" cry two people. "Yes, he's sitting under that tree."

---

1. **Federal**：美國南北戰爭時期聯邦黨人。
2. **Democrat**：民主黨黨員。
3. **Congressman**：美國國會議員，代表一個州。

Rip Van Winkle

Rip looks and sees a young man with dirty old clothes.

"Is that me? Then WHO am I?" asks Rip.

They look at him and say, "He must be crazy." [1]

Then a young woman carefully looks at Rip.

Rip asks, "What is your name?"

"Judith Gardiner."

"What is your father's name?"

"Rip Van Winkle! Twenty years ago he went away and he never returned," she says.

"Where is your mother?" Rip asks.

"She is dead."

Rip embraces [2] his daughter and says, "I'm your father. I'm Rip Van Winkle. I'm old now."

He turns to the crowd [3] and says, "Does anyone remember Rip Van Winkle?"

A very old woman says, "Yes, this man IS Rip Van Winkle senior. [4] I remember him."

Rip tells his incredible [5] story. The people of the village can't believe his story is true. They ask old Peter Vanderdonk. He is the oldest man in the village. He knows all the legends and stories of the Catskill Mountains.

Peter Vanderdonk says, "Mysterious things happen in the Catskill Mountains. There are strange spirits living there. People say that every twenty years Henry Hudson [6] returns to see

---

1. **crazy**：發瘋的。
2. **embraces**：擁抱。
3. **crowd**：人羣。
4. **senior**：置於姓名後表示同名父子中的父或同姓二人之中較年長者。
5. **incredible**：難以置信的。
6. **Henry Hudson**：英國航海家、探險家，於十七世紀初發現哈德遜河。

The Spirits of the Catskill Mountains

his river and that he and his men play ninepins in those mountains."

It is late and the people of the village go home. Rip goes to live with his daughter and her family, and his son visits him every day. He is very similar to his father. He does everyone's work but not his own.

Rip makes some new friends. Everyone likes him. He doesn't really understand the American Revolutionary War [1] or the new nation called the United States of America. But he is a happy man now. He sits outside the Union Hotel and tells stories about the past. He also tells travelers about his twenty-year- long sleep. Some people believe him but some don't.

---

1. **the American Revolutionary War**：美國獨立戰爭（1775-1783）。

UNDERSTANDING THE TEXT

KET

1 **Are these sentences "Right" (A) or "Wrong" (B)? If there is not enough information, choose "Doesn't say" (C). Circle the correct answers.**

1. The Union Hotel is blue and red.

   A Right   **B** Wrong   **C** Doesn't say

2. A thin man is talking about King George.

   A Right   **B** Wrong   **C** Doesn't say

3. Rip doesn't understand what the man is saying.

   A Right   **B** Wrong   **C** Doesn't say

4. The people of the village think Rip is a spy.

   A Right   **B** Wrong   **C** Doesn't say

5. Rip looks for his friends but they are all dead.

   A Right   **B** Wrong   **C** Doesn't say

6. Rip is very confused and cries, "Does anyone know Rip Van Winkle?"

   A Right   **B** Wrong   **C** Doesn't say

7. Judith Gardiner has three children.

   A Right   **B** Wrong   **C** Doesn't say

8. The people of the village can't believe Rip's incredible story.

   A Right   **B** Wrong   **C** Doesn't say

9. Derrick Bummel knows all the legends and stories of the Catskill Mountains.

   A Right   **B** Wrong   **C** Doesn't say

10. Peter Vanderdonk tells the strange story of Henry Hudson and his men.

    A Right   **B** Wrong   **C** Doesn't say

11. Rip goes to live with his daughter and is a happy man.

    A Right   **B** Wrong   **C** Doesn't say

After his long sleep, Rip wants to know more about the United States. So, he asks his friend Congressman Derrick Bummel. Listen to the interview between Rip and Congressman Bummel. Then tick (✓) the correct answers.

1. The United States of America is a

   A ☐ democracy.

   B ☐ monarchy.

   C ☐ British colony.

2. Who chooses the president of the United States?

   A ☐ Congressmen.

   B ☐ The King.

   C ☐ The people.

3. Who is the first American president?

   A ☐ Derrick Bummel.

   B ☐ George Washington.

   C ☐ King George.

4. George Washington is

   A ☐ a congressman.

   B ☐ an old innkeeper.

   C ☐ a hero of the American Revolution.

5. In a democracy everyone is equal and

   A ☐ rich.

   B ☐ free.

   C ☐ brave.

Which notice (a - f) says this (1 - 5)?

1. ☐ We sell food.

2. ☐ We don't close for lunch.

3. ☐ We don't work on Sundays.

4. ☐ We make cakes on Saturday.

5. ☐ We are friendly.

 Your opinion:

Do you like the story?

Why? I like the story because it ....

I don't like the story ....

What part of the story is your favorite? ...I like the part when...

....

THE AMERICAN REVOLUTION

1 A journalist from the *Boston Gazette* interviews General George Washington. Listen to the interview, then tick (✓) the correct answers.

1. How many American colonies are there?
- **A** ☐ 30.
- **B** ☐ 3.
- **C** ☐ 13.

2. Who is their leader?
- **A** ☐ King George.
- **B** ☐ George Washington.
- **C** ☐ The journalist.

3. The colonies want to
- **A** ☐ become an independent nation.
- **B** ☐ pay taxes to King George.
- **C** ☐ obey British laws.

4. The American uniforms are
- **A** ☐ blue and red.
- **B** ☐ red and white.
- **C** ☐ blue and white.

5. The Declaration of Independence is
- **A** ☐ a Boston newspaper.
- **B** ☐ an important document.
- **C** ☐ a British tax.

Now check your answers by reading the interview on page 42.

INTERVIEW

A journalist from the *Boston Gazette* interviews General George Washington.

Journalist: Hello, General Washington. What's happening in Boston?

Washington: Well, there are 13 American colonies protesting against [1] Great Britain and I'm their leader.

Journalist: Why are you protesting?

Washington: We don't want to obey British laws and we don't want to pay taxes to King George.

Journalist: What do you want?

Washington: We want to become an independent nation.

Journalist: What are you going to do?

Washington: We are going to fight the British until we win! We have muskets, [2] swords, horses and many courageous men. We're called "Yankees". But not all colonists are Yankees. Some are redcoats [3] and fight with the British.

Journalist: What color are your uniforms? [4]

Washington: Our uniforms are blue and white. The British wear red jackets. On July 4 my friends and I are going to sign the Declaration of Independence. [5]

Journalist: What's that?

Washington: It's a very important document that says the thirteen colonies are free and independent states.

Journalist: Good luck to you and your men!

Washington: Thank you.

---

1. **protesting against**：反對。
2. **muskets**：火槍。
3. **redcoats**：美國獨立戰爭時期的英兵（因其所穿紅制服得名）。
4. **uniforms**：制服。
5. **the Declaration of Independence**：（美國）《獨立宣言》。

Today Americans still celebrate July 4 or Independence Day. It is a national holiday.

PROJECT

Look for more information about the Fourth of July on the web. Find out how Americans celebrate this national holiday.

Use one of the following search engines:
http://www.yahoo.com
http://www.infoseek.com
http://www.altavista.com

Then type in July 4, 1776 or Fourth of July.
HAVE FUN!

EXIT TEST

1 **Who are they? Match the numbers with the letters.**

Who...

1. ☐ nags her husband?
2. ☐ is Rip's dog?
3. ☐ is the innkeeper?
4. ☐ is the oldest man in the village?
5. ☐ is the school teacher?
6. ☐ is Rip's daughter?

a. Wolf
b. Peter Vanderdonk
c. Derrick Bummel
d. Judith Gardiner
e. Nicholas Vedder
f. Mrs Van Winkle

2 **Are the following sentences true (T) or false (F)? Correct the false ones.**

	T	F
1. Rip Van Winkle lives in a village at the foot of the Catskill Mountains.	☐	☐
2. Rip likes working.	☐	☐
3. Mrs Van Winkle is always kind to her husband.	☐	☐
4. During the summer Rip sits outside the inn with his friends.	☐	☐
5. Rip meets Derrick Bummel in the mountains.	☐	☐
6. The men are playing a game of tennis.	☐	☐
7. The drink in the barrel is very strong.	☐	☐
8. Rip wakes up after twenty years.	☐	☐
9. Soon he finds his old friends and his family.	☐	☐
10. Every twenty years Henry Hudson returns to see his river.	☐	☐
11. Rip tells travelers about his long sleep.	☐	☐

The Legend
of Sleepy Hollow

Here are some words from the story. Do you know them?

ghost, spirit
鬼魅，幽靈

witch
巫師

heart
心臟

pumpkin
南瓜

handkerchief
手帕

bridge
橋

saddle
馬鞍

sword
劍

shadow
影子

musket
火槍

CHAPTER ONE

The Headless Horseman

Greensburgh is a village on the Hudson River. Near this village there is a silent valley called Sleepy Hollow. It is the quietest place in the world. The inhabitants [1] are always a bit sleepy there. Some people say there are bizarre spirits in this valley. Others say there are strange voices in the air.

The people of Greensburgh believe there is a ghost in Sleepy Hollow. It is the ghost of a headless [2] man on a big horse. Legend says it is the ghost of a soldier of the American Revolutionary War.

---

1. **inhabitants** : 居民。
2. **headless** : 無頭的。

The Legend of Sleepy Hollow

His body is buried [1] in the cemetery [2] near the church.

Every night he rides his horse and looks for his head. He rides in the valley, near the river, in the forest and on the roads. He always returns to the cemetery before daylight. Everyone calls him the Headless Horseman of Sleepy Hollow.

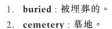

1. **buried**：被埋葬的。
2. **cemetery**：墓地。

UNDERSTANDING THE TEXT

1 **Are these sentences "Right" (A) or "Wrong" (B). If there is not enough information to answer, choose "Doesn't say" (C). Circle the correct answers.**

1. Sleepy Hollow is the quietest place in the world.
   A Right   **B** Wrong   **C** Doesn't say

2. People say there are witches in the valley.
   A Right   **B** Wrong   **C** Doesn't say

3. The inhabitants of Sleepy Hollow are always a bit sleepy.
   A Right   **B** Wrong   **C** Doesn't say

4. In Greensburgh there is a big church.
   A Right   **B** Wrong   **C** Doesn't say

5. The people of Greensburgh don't believe in ghosts.
   A Right   **B** Wrong   **C** Doesn't say

6. The ghost of Sleepy Hollow is a headless man and he rides a big horse.
   A Right   **B** Wrong   **C** Doesn't say

7. His body is buried in the forest.
   A Right   **B** Wrong   **C** Doesn't say

2 **Match the words with their synonyms. Write the correct letters in the boxes.**

1. ☐ silent   **a.** strange
2. ☐ enchanted   **b.** spirit
3. ☐ bizarre   **c.** tranquil
4. ☐ ghost   **d.** tired
5. ☐ sleepy   **e.** magic

Use some of the words above to write a sentence describing Sleepy Hollow, then write to a friend describing your own town.

Sleepy Hollow is ....

My town is ....

CHAPTER TWO

Ichabod Crane

I chabod Crane lives in Sleepy Hollow. He is a school teacher and comes from the state of Connecticut. He is tall and very thin. His legs and arms are very long. His feet are very big. His head is small with big ears, green eyes and a long nose. His clothes are always too big for him. Ichabod is always hungry and loves to eat.

Ichabod is a good but strict teacher. His school is in the middle of the country. It has only one big room for all the pupils.

After school Ichabod plays ball with the older boys and on his free afternoons he walks home with some of the younger ones.

The Legend of Sleepy Hollow

He always chooses the boys with pretty sisters or generous[1] mothers.

Ichabod does not have much money, so every week he lives with a different family. He eats and sleeps at their house and keeps them company.[2] When he changes family he puts all his things in a big handkerchief and carries it with him.

The farmers like him because he helps them with their work and the mothers like him because he plays with the small children.

Ichabod also teaches singing. On Sundays he is the director of the church choir.[3] He is very proud of this.

The school teacher is an important person among farmers and their families. Everyone respects him because he is educated and the women consider him a gentleman. Ichabod likes gossiping[4] and being the center of attention. He also likes eating all the good food the women prepare for him.

Ichabod loves reading books. One of his favorite books is

1. **generous**：慷慨的。
2. **keeps them company**：陪伴他們。
3. **church choir**：教會的詩班。
4. **gossiping**：說別人閒話。

Ichabod Crane

History of New England Witchcraft. [1] Ichabod believes the stories in this book. They are about ghosts and witches.

During the winter Ichabod often visits the old women. They tell him terrible stories about ghosts, haunted [2] houses, haunted trees and the Headless Horseman. When he returns home at night the sounds of the forest scare him. He thinks there are ghosts everywhere. So he sings church songs. Ichabod always sings these songs when he is afraid.

---

1. **witchcraft**：巫術。
2. **haunted**：鬼魂出沒的。

UNDERSTANDING THE TEXT

1 **Complete the questions with the correct letters.**

1. Where does Ichabod Crane live? . . . .
2. What color are his eyes? . . . .
3. Who is the school teacher? . . . .
4. Where is Ichabod Crane's school? . . . .
5. Who are his pupils? . . . .
6. What does Ichabod do after school? . . . .
7. Why do the farmers like him? . . . .
8. Who does Ichabod visit during the winter? . . . .
9. What does Ichabod do on Sundays? . . . .
10. What is the name of his favorite book? . . . .

a. In the country.
b. He helps them with their work.
c. In Sleepy Hollow.
d. The old women.
e. Ichabod Crane.
f. Farmers' children.
g. He plays ball with the older boys.
h. He is the director of the church choir.
i. They are green.
j. *History of New England Witchcraft.*

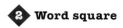

2 Word square

There are seven parts of the body in the Word Square repeated three times. Can you find all twenty-one of them?

H	E	A	D	X	H	A	C	V	H	U	M	L
H	F	L	E	G	S	Y	X	U	N	O	S	E
D	R	A	A	S	E	O	G	H	T	K	I	G
E	F	T	R	K	W	N	F	E	E	T	Y	S
Y	G	E	S	U	A	O	E	A	R	S	A	K
E	N	S	E	A	R	S	E	D	R	C	R	H
S	O	M	K	T	M	E	T	K	Y	M	M	E
Z	S	I	T	W	S	D	G	E	Y	E	S	A
L	E	G	S	A	U	E	Y	E	S	T	P	D

Now use the words to fill in the gaps and describe Ichabod Crane.

His .... and .... are very long. His ....
are very big. His .... is small with big ...., green
.... and a long ....

You want to tell your best friend about a new boy/girl friend. Write a description of him/her.

....
....

Now describe your favorite actor/actress.

....
....

CHAPTER THREE

Katrina Van Tassel

Katrina Van Tassel is one of Ichabod's singing students. She is the daughter of a rich farmer. She is eighteen years old and is very beautiful. She has rosy cheeks [1] and a lovely smile. She always wears pretty clothes.

Katrina's father has a big home with beautiful furniture. [2] He also has a fine farm, many animals and lots of good food. Ichabod likes her father's riches. [3] He also likes Katrina very much and wants to marry her.

"Oh, how I like Katrina!" he thinks. "How can I win her heart?"

---

1. **cheeks** :
2. **furniture** : 家具。
3. **riches** : 財富。

Katrina Van Tassel

Katrina has many admirers. [1] They often visit the Van Tassel house. Who does Katrina like? No one knows. One day she is nice to one young man and the next day she is nice to another.

---

1. **admirers** : 追求者。

The Legend of Sleepy Hollow

Brom Bones is the hero of Sleepy Hollow because he is tall, strong and handsome. He has broad [1] shoulders. He is a bit rough [2] but always cheerful. He is an excellent horseback rider and a courageous fighter. Everyone likes Brom but sometimes he is mischievous [3] and plays tricks. He is Ichabod's big rival. [4] Brom often visits the Van Tassel house and courts [5] Katrina. When his horse is outside the Van Tassel house, other admirers stay away.

Ichabod also visits Katrina often because he is her singing teacher. He and Katrina sit together under the big tree singing and talking. Or they walk together along the river in the evenings. Ichabod slowly wins Katrina's heart.

Brom is very angry because he wants to marry beautiful Katrina. He decides to fight Ichabod. But Ichabod doesn't want to. "Brom is big and strong. I cannot fight him and win," Ichabod thinks.

Brom tries to make Ichabod angry. One night he goes into his school and he throws the tables and chairs about. The next day Ichabod goes to school and says, "Oh, no! This is where the witches meet at night!"

Brom is very mischievous and teaches his dog to make terrible noises. Then he tells Katrina, "Listen to my dog. His singing teacher is Ichabod Crane!"

But nothing can make Ichabod fight so Brom is furious with him. "I must find another way," Brom thinks.

---

1. **broad**：寬闊的。
2. **rough**：粗魯的。
3. **mischievous**：頑皮的，愛搗亂的。
4. **rival**：競爭對手。
5. **courts**：（此處指）追求。

UNDERSTANDING THE TEXT

1 Choose the correct words to complete the sentences. Tick (✔) A, B or C.

1. Katrina Van Tassel is young and
   - **A** ☐ fat.
   - **B** ☐ beautiful.
   - **C** ☐ ugly.

2. She is the daughter of
   - **A** ☐ the singing teacher.
   - **B** ☐ a poor farmer.
   - **C** ☐ a rich farmer.

3. Ichabod likes Katrina very much and wants to
   - **A** ☐ win her heart.
   - **B** ☐ dance with her.
   - **C** ☐ give her a book.

4. Katrina has many
   - **A** ☐ horses.
   - **B** ☐ admirers.
   - **C** ☐ clothes.

5. Brom Bones is
   - **A** ☐ a rich farmer.
   - **B** ☐ a soldier.
   - **C** ☐ Ichabod's rival.

6. Brom wants to
   - **A** ☐ fight Ichabod.
   - **B** ☐ learn to sing.
   - **C** ☐ fight the Headless Horseman.

2 **What does Katrina say to Ichabod?**
Complete the conversation. Put the correct letters
in the blanks.

Ichabod: Hello, Katrina!

Katrina:.........................

Ichabod: How are you?

Katrina:....................

Ichabod: I'm happy to see you.

Katrina:.................

Ichabod: Is this a new dress?

Katrina: ....

Ichabod: It's lovely!

Katrina: ....

Ichabod: What time is dinner?

Katrina:..........

Ichabod: I'm very hungry.

Katrina: ....

Ichabod: Wonderful!

a. There's a lot of good food this evening.

b. At half past six.

c. Yes, it is.

d. Good evening, Ichabod.

e. I am too.

f. Oh, thank you.

g. Very well, thank you.

h. Good Night!

3 Listen to this dictation twice and fill in the missing words.

Katrina .... a beautiful girl. She ....
in a .... house on a big .... Her
father has a lot of .... and her mother
.... a lot of good .... She has many
.... and she is .... to all of them.

Brom often .... Katrina because he ....
her very much. He is a tall, ... man.

Ichabod is .... singing teacher. He also
.... Katrina often. Ichabod .... her
.... farm and her .... food a lot. He
has very long .... and ....
He .... like a crane.

Who .... Katrina prefer?

4 Unscramble the letters and make the question. Then complete this word puzzle.

<u>W</u> <u>h</u> <u>o</u> _ _ _ _   _ _ _   _ _ _ _ ?

a. You wear them every day _ _ _ _ <u>h</u> _ _

b. Very pretty _ _ _ _ _ _ _ _ <u>l</u>

c. Not always polite or gentle _ <u>o</u> _ _ _

d. Very angry _ _ _ _ <u>o</u> _ <u>s</u>

e. Playful and sometimes causes trouble
_ _ _ _ <u>h</u> _ <u>e</u> <u>v</u> _ _ _

f. To make afraid <u>s</u> _ _ _ _ <u>e</u>

g. A cold season <u>w</u> _ _ _ <u>e</u> _

h. Wide, big shoulders _ _ <u>o</u> _ <u>d</u>

61

NEW YORK IN THE 1800S

Mr Van Tassel writes a letter to his brother in Holland. Read it and learn about New York in the 1800s.

Sleepy Hollow, October 28, 1820

Dear Peter,

Here I am in Sleepy Hollow in the state of New York. It is a very quiet place and it is full of farmers. My farm is very big. I have a lot of animals, vegetables and fruit trees but I must work very hard.

Every week I go to the city of New York to sell my products. [1] I sell: apples, pears, pumpkins, vegetables, milk, butter, cheese and eggs. I put them all on a boat and then go down the Hudson River to New York City. The people there like good food from the country.

Of course, I take the food to a big market because New York City is really big. About 150,000 people live there. Everyone moves about very quickly. The streets are wide with many horse-drawn carriages. [2] I like looking at the big shops and tall buildings especially in the business area. It is called Wall Street.

New York has an important harbor [3] with many factories near it. The ships in the harbor come from other American states or from Europe. I am always surprised by the big number of immigrants [4] from Ireland, Germany and other countries.

There are not many Indians on the streets but the ones you can see still wear their native dress. [5]

I hope you can come to visit me soon and see this cosmopolitan [6] city.

Your brother,
Balthus

1. **products** : 產品。
2. **horse-drawn carriages** : 馬車。
3. **harbor** : 港口。
4. **immigrants** : 外來移民。
5. **native dress** : 本國服裝。
6. **cosmopolitan** : 國際性的。

1 **Mr Van Tassel sells fruits, vegetables and dairy [1] products at the market but no meat. Circle all the food words and list them under each heading.**

M	E	A	U	Q	B	O	C	S	B	G	D	T
B	F	P	Z	E	G	G	S	H	U	D	C	G
C	X	P	U	M	R	O	U	A	T	R	O	V
S	H	L	Y	M	Y	N	E	M	T	X	I	P
T	L	E	D	U	P	E	H	P	E	A	R	S
V	I	S	E	I	S	K	L	C	R	B	D	H
U	V	I	P	S	R	D	I	T	M	W	M	Z
Z	P	E	Q	O	E	A	S	N	F	C	I	K
O	N	G	B	D	B	S	O	W	S	K	L	L
A	V	E	G	E	T	A	B	L	E	S	K	N

Vegetables/Fruits

....
....
....
....
....

Dairy

....
....
....
....
....

PROJECT

Look for more information about New York on the web.
Find out what you can do and see there.

Use one of the following search engines:
http://www.yahoo.com
http://www.infoseek.com
http://www.altavista.com

Then type in New York.
HAVE FUN!

1. **dairy** : 牛奶製的。

CHAPTER FOUR

The Invitation

One autumn afternoon Ichabod is teaching at school. A messenger [1] brings him a letter. It is an invitation to dinner at the Van Tassels' house. Ichabod is happy and excited. His pupils are happy too because they finish school early.

He puts on his best clothes: an old dark suit. Then he looks at himself in a broken mirror. "I want to be handsome for Katrina," he thinks. "This is an important occasion for me."

Ichabod is staying at Van Ripper's farm this week. So, after school Ichabod returns to the farm and asks Van Ripper, "Can I use your horse this evening?"

"Yes, you can use Gunpowder. [2] He is old now and not very handsome. He has only one eye. But he is still a very good horse," says Van Ripper.

---

1. **messenger** : 信差。

2. **Gunpowder** : 火藥。

The Invitation

"Thank you, Mr Van Ripper," says Ichabod. He looks at Gunpowder and thinks, "He is certainly an old horse, but he is better than nothing." Then he looks carefully into Gunpowder's only eye and thinks, "I see the devil in his eye! He is still a lively horse."

Ichabod and Gunpowder look funny together: Ichabod in his old suit and Gunpowder with his one eye. They leave Van Ripper's farm, cross the forest and ride along the Hudson River.

Autumn is a beautiful season in Sleepy Hollow. There are many different colors: green, red, yellow, orange, purple [1] and brown. It is sunset and the sky is the color of gold. Along the road Ichabod sees orange pumpkins and apple trees. All he can think about are the delicious [2] pies [3] and cakes.

---

1. **purple** : 紫色。
2. **delicious** : 美味的。
3. **pies** :

The Legend of Sleepy Hollow

Ichabod finally arrives at the Van Tassels' house. All the farmers are there with their families. They are wearing their best clothes. Everyone is talking and laughing. It is a happy evening.

Brom is there too. Everyone admires his horse, Daredevil. [1] Daredevil is a big, black horse. Only Brom can ride him because he is a dangerous animal. Brom is very proud of Daredevil.

Ichabod is delighted [2] when he sees the delicious food on the Van Tassels' table. His big eyes move from one dish to the next. There are all kinds of cakes: sweet cakes, ginger cakes [3] and honey cakes. There are also pies, meat, fish, chicken, vegetables, fruit and lots of milk.

"This is a royal banquet!" [4] he thinks. "When I marry Katrina I can have all this food, this beautiful home and lots of money. I can laugh at Van Ripper and his old horse. I can laugh at everyone." Ichabod eats and drinks and is very happy.

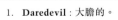

---

1. **Daredevil**：大膽的。
2. **delighted**：高興的。
3. **ginger cakes**：薑餅。
4. **royal banquet**：盛宴。

UNDERSTANDING THE TEXT

1 **Read the paragraph and write the best word A, B or C in each space. The first one is done for you.**

A messenger ¹......C...... Ichabod an invitation. He is happy ².............. excited. He puts ³.............. his best clothes. Ichabod ⁴.............. Van Ripper's old horse Gunpowder. Gunpowder is very old and ⁵.............. only one eye.

Ichabod and Gunpowder ride ⁶.............. the Hudson River.

Ichabod arrives ⁷.............. the Van Tassel house ⁸.............. the evening. There ⁹.............. a lot of delicious food ¹⁰.............. the table. Ichabod's eyes move ¹¹.............. one dish ¹².............. the next.

1. A to bring	**B** bring	**C** brings
2. A and	**B** but	**C** or
3. A at	**B** on	**C** in
4. A uses	**B** use	**C** used
5. A is	**B** have	**C** has
6. A along	**B** in	**C** at
7. A in	**B** at	**C** to
8. A by	**B** at	**C** in
9. A are	**B** is	**C** be
10. A in	**B** from	**C** on
11. A from	**B** to	**C** at
12. A at	**B** on	**C** to

2 **What does Katrina write to Ichabod? Listen to the letter twice and fill in the gaps.**

Dear Ichabod,

There's a Halloween party .... Please come to my .... for dinner at ....
I am .... a chocolate cake. My .... is preparing a ....
My .... is preparing a ....

After dinner we can .... in the ....
You can meet my five .... from New York.
At .... we can .... and
.... ! Please wear your ....

Your friend,
Katrina

3 **Look at the picture below. Ichabod is looking at the good food on the banquet table. Label the food using some of the words in the box.**

chicken   bread   fish   meat   cake   pear
pumpkin   apple   pie   cakes   potatoes   grapes

CRISPY [1] CHOCOLATE CAKES

Do you like chocolate? Try this yummy [2] recipe.

You need:

a large bowl, a large saucepan, a wooden spoon, a teaspoon, 12 paper cake cases, a baking sheet

Ingredients:

225 g milk chocolate, 85 g cornflakes [3] or puffed rice [4]

1. Melt the chocolate in a bowl.

2. Mix the breakfast cereal [5] with the chocolate.

3. Put the mixture into the paper cases. Put them on a baking sheet and let them cool.

The chocolate crispy cakes take about 1 hour to set.

1. **crispy**：脆的。
2. **yummy**：美味的。
3. **cornflakes**：玉米片。
4. **puffed rice**：膨脹的米。
5. **cereal**：穀類食品。

CHAPTER FIVE

The Ghosts of Sleepy Hollow

After dinner a small orchestra [1] plays music. Ichabod dances with Katrina. He says sweet things to her. Katrina smiles at Ichabod and he smiles at her.

Brom isn't dancing. He is sitting in a corner. He is watching Katrina and Ichabod. He is angry and jealous. [2] "Ichabod is a good dancer but I can't dance. What can I do to win Katrina?"

After the dance everyone goes outside and sits under the trees. They talk about the American Revolutionary War.

1. **orchestra**：管弦樂隊。
2. **jealous**：妒忌的。

70

The Ghosts
of Sleepy Hollow

"When I look at the river I remember the battles between the British and the Americans. I fought alone against the British with my small sword!" says an old farmer.

A young farmer says, "When the British saw my new musket they ran away!"

"We are all heroes [1] of the American Revolutionary War," says another man.

They also talk about ghosts.

---

1. **heroes** : 英雄。

The Legend of Sleepy Hollow

"Raven Rock is haunted. There's the ghost of a woman in white. I always see her ghost before a storm. She's scary!" [1] says a young woman.

"When I pass by the big tree near the church I hear terrible cries. It's the ghost of Major André," says a boy.

"The Headless Horseman is the most horrible. [2] He gallops [3] on his big black horse and follows you. Sometimes he takes you away! When there is a full moon you can see him clearly!" says a tall man. Ichabod listens to these stories and believes them.

Soon the farmers and their families start going home. Ichabod wants to stay and talk to Katrina. "This is the right moment. I must ask Katrina to marry me," he thinks.

No one knows what happens between Katrina and Ichabod. But he is very sad when he leaves the Van Tassel house. Does Katrina prefer [4] Brom?

Ichabod gets on Gunpowder and starts riding to Van Ripper's farm. It is a very dark night. Clouds cover the stars. Ichabod thinks about the ghosts of Sleepy Hollow. He hears a strange noise in the forest. Is it the voice of a ghost or is it the wind? Something is moving in the sky. Is it a witch? Is it the ghost of the woman in white?

"What a mysterious night!" Ichabod thinks. He is lonely and sad. He rides very quickly.

He comes to a brook [5] with a small bridge. It is very difficult

---

1. **scary**：駭人的。
2. **horrible**：可怕的。
3. **gallops**：騎馬奔馳。
4. **prefer**：（兩者比較）更喜歡。
5. **brook**：小溪。

The Ghosts of Sleepy Hollow

to cross the bridge. On the other side of the brook Ichabod sees some enormous [1] trees. It is dark and he is afraid.

"Major André's ghost lives here," he thinks. "I must cross this bridge quickly."

Ichabod kicks Gunpowder and says, "Come on, Gunpowder. Cross the bridge quickly!" But the old horse does not move. Ichabod kicks Gunpowder again and again. Finally the horse moves but then he stops suddenly. Ichabod nearly falls off. He is very nervous.

"What is happening to me tonight?" he thinks.

Then he hears a strange sound on the other side of the brook. In the shadow of the trees he sees something big and black.

"Oh, no, something is waiting for me. It isn't moving. It is waiting for me. I cannot go back. What must I do?" he thinks. His heart is beating fast. His hair is standing up. His mouth is dry and his hands are very cold.

"Who are you?" Ichabod asks.

There is no answer.

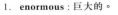

1. **enormous**：巨大的。

The Legend of Sleepy Hollow

"Who are you?" he asks again. His teeth [1] are chattering. [2]
There is still no answer.

Ichabod closes his eyes and starts singing a church song.

At that moment the big, black thing moves again. It jumps out into the road and Ichabod opens his eyes. The night is dark but he can see a horseman on a big, black horse.

1. **teeth**：（單數形式為 tooth）牙齒。
2. **chattering**：因害怕而牙齒咯咯作響。

74

UNDERSTANDING THE TEXT

 1 **Choose the correct words to complete the sentences. Tick (✔) A, B or C.**

1. Brom is sitting in a corner because he
A ☐ is eating.
B ☐ can't dance.
C ☐ is talking to Katrina.

2. Some people say the Headless Horseman
A ☐ stays at Raven Rock.
B ☐ is Major André.
C ☐ gallops all night on a black horse.

3. When Ichabod leaves the Van Tassel house
A ☐ he is very sad.
B ☐ Katrina walks with him.
C ☐ he sees Brom and Daredevil.

4. In the forest Ichabod hears
A ☐ the voice of a ghost.
B ☐ a strange noise.
C ☐ the wind.

5. Ichabod's heart beats fast and his hair stands up because
A ☐ he is afraid.
B ☐ there is a storm in the forest.
C ☐ he is very hungry.

6. In the shadow of the trees Ichabod
A ☐ hears Major André's voice.
B ☐ sees something big and black.
C ☐ sees an old witch.

❷ Have fun with this crossword puzzle!

ACROSS

2.

4.

7.

8.

10.

12.

14. not always polite or gentle

16.

17. They are in your mouth

18. without a head

DOWN

1.

3. Van Ripper's old horse

5. small hotel or tavern

6. a navigator and explorer

9.

11. wide, big

13.

15.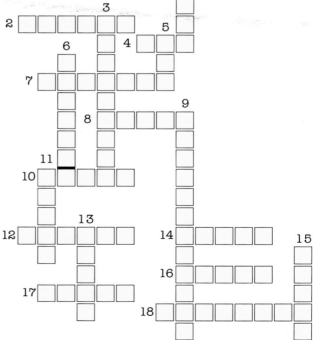

CHAPTER SIX

The Night Ride

chabod forces [1] the horse to gallop. The horseman follows him. Ichabod thinks about the horrible stories of the Headless Horseman. Gunpowder gallops fast but the horseman gallops fast too. Ichabod wants to sing a church song but his mouth is dry.

They gallop up a hill. Against the night sky Ichabod sees the horseman. He is headless! He is carrying his head on his saddle. Ichabod is terrified. [2] He hits Gunpowder again and again. He wants the old horse to gallop fast. But the Headless Horseman follows him with his big, black horse.

The road to Van Ripper's farm is on the left. But Gunpowder turns right. This road goes to the church bridge and the church cemetery.

Ichabod and Gunpowder gallop down the hill. Suddenly Ichabod's saddle falls off, but the horse does not stop. "Oh, no!"

1. **forces** : 強迫。　　　　　　　　　　2. **terrified** : 害怕的。

The Legend of Sleepy Hollow

thinks Ichabod. "That is Van Ripper's best saddle!"

Ichabod clasps Gunpowder's neck. He tries to stay on the horse. The Headless Horseman is behind him. Ichabod is very scared. His teeth are chattering.

"The church bridge is near here," he thinks. "Ghosts usually disappear after the church bridge. I must cross the bridge, then I am safe!"

---

1. **clasps**：抱緊。

The Night Ride

Ichabod and Gunpowder gallop across the church bridge. Ichabod looks back and the Headless Horseman is still there! He has a big round object in his hands. He throws it at Ichabod.

"It's his head!" cries Ichabod. The big head hits Ichabod and there is a loud thud. [1] Ichabod falls to the ground and Gunpowder gallops away.

---

1. **thud** : 砰的一聲，重物墜落的聲音。

The Legend of Sleepy Hollow

The next morning Gunpowder arrives at Van Ripper's farm but Ichabod isn't there for breakfast. Ichabod isn't at school and he isn't at Van Ripper's farm.

"Where is Ichabod," asks Van Ripper, "and where is my best saddle?"

Van Ripper and his friends look for Ichabod everywhere. They find the saddle on the road near the church. They find Ichabod's hat and a broken pumpkin near the river. But they can't find Ichabod's body.

"It's the Headless Horseman. Poor Ichabod Crane!" says Van Ripper.

"Yes, it's the Headless Horseman. Oh poor Ichabod!" say his friends.

Years later an old farmer returning from New York says to the people of Sleepy Hollow, "Ichabod Crane is alive [1] and well! Now he is a Congressman. He left Sleepy Hollow because Katrina wanted to marry Brom Bones."

Katrina hearing this asks her husband Brom, "What do you know about Ichabod Crane and the pumpkin?"

Brom laughs loudly.

At Sleepy Hollow people say that only Brom knows the true story of Ichabod Crane and the Headless Horseman. It is Brom's secret.

Now, on winter evenings the old women tell the terrible story of Ichabod Crane and the Headless Horseman.

1. **alive** : 活着。

UNDERSTANDING THE TEXT

KET

1 **Are these sentences "Right" (A) or "Wrong" (B)? If there is not enough information to answer, choose "Doesn't say" (C). Circle the correct answers.**

1. It is Halloween night.

   A Right   **B** Wrong   **C** Doesn't say

2. Ichabod gallops fast and the Headless Horseman follows him.

   A Right   **B** Wrong   **C** Doesn't say

3. The Headless Horseman carries his head on his saddle.

   A Right   **B** Wrong   **C** Doesn't say

4. He has a big, white horse.

   A Right   **B** Wrong   **C** Doesn't say

5. Ghosts appear after the church bridge.

   A Right   **B** Wrong   **C** Doesn't say

6. The Headless Horseman throws his head at Ichabod.

   A Right   **B** Wrong   **C** Doesn't say

7. Ichabod falls to the ground and starts singing a church song.

   A Right   **B** Wrong   **C** Doesn't say

8. The next day Van Ripper and his friends find Ichabod's hat and a broken pumpkin.

   A Right   **B** Wrong   **C** Doesn't say

9. Brom becomes a Congressman.

   A Right   **B** Wrong   **C** Doesn't say

10. Only Brom knows the true story of Ichabod Crane and the Headless Horseman.

   A Right   **B** Wrong   **C** Doesn't say

81

a. Who is the Headless Horseman?

....

b. What is Brom's secret?

....

c. Who is your favourite character?

....

d. Do you like this story?

Why? ....

....

*Felix Darley's original illustration of Ichabod
Crane for the 1848 edition of the story.*

EXIT TEST

1 Describe these characters. Find the correct words in the box below and write them under each picture.

KATRINA

....
....
....
....
....
....
....
....

ICHABOD

....
....
....
....
....
....
....
....

BROM

....
....
....
....
....
....
....
....

mischievous	young hero	eighteen years old

very thin   beautiful   likes gossiping
courageous fighter   always hungry   strong   rich
has a lovely smile   plays tricks   tall
school teacher   young   big   singing teacher
excellent horseback rider   educated   likes reading
wears pretty dresses   has rosy cheeks

2 **Unscramble the sentences and write them on the dotted lines. You will have a summary of the story!**

a. is a school teacher   Ichabod Crane   and a singing teacher
....

b. and eating   Ichabod likes   reading books   good food
....

c. the beautiful daughter   Katrina is   farmer   of a rich
....

d. want to marry   Both Ichabod and Brom   Katrina
....

e. are   Ichabod and   rivals   Brom
....

f. goes to dinner   Van Tassels' house   Ichabod   at the
....

g. the Headless Horseman   in the forest   Ichabod meets
....

h. follows   The Headless Horseman   Ichabod
....

i. throws   at Ichabod   The Headless Horseman   his head
....

j. can find   the next morning   Ichabod   No one
....

k. and keeps   Brom   marries Katrina   his secret
....

Rip Van Winkle and The Legend of Sleepy Hollow

KEY TO THE EXERCISES AND EXIT TEST

RIP VAN WINKLE

CHAPTER ONE

Page 19 Exercise 1
1. A   2. A   3. B   4. C   5. A   6. B
7. A   8. B   9. C   10. B

Page 20 Exercise 2
1. B   2. A   3. C   4. B   5. A   6. C
7. B   8. B

CHAPTER TWO

Page 24 Exercise 1
1. B   2. C   3. A   4. C   5. A
6. B   7. C

Page 25 Exercise 2
1. c   2. d   3. f   4. e
5. g   6. a   7. b

Page 26 Exercise 3
a. up

b. on
c. around
d. on
e. toward
f. into
g. of
h. from

Page 26 Exercise 4
The mysterious drink in the barrel is magic.

CHAPTER THREE

Page 30 Exercise 1
1. B   2. C   3. A   4. C   5. B
6. B   7. A   8. A   9. C   10. C   11. A
12. B   13. A   14. C

Page 31 Exercise 2
1. d   2. e   3. a   4. c   5. b

85

Page 31 Exercise 3

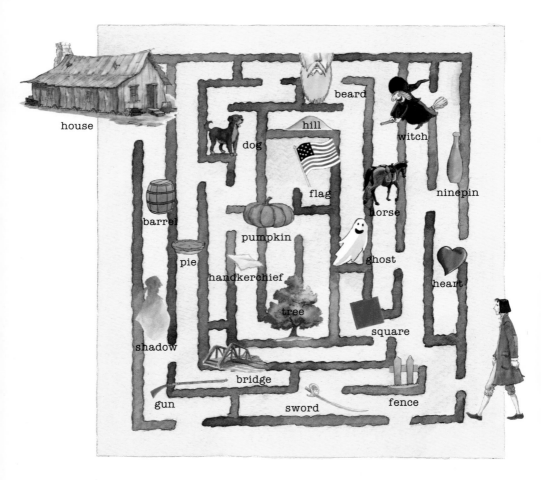

There are seven words with the letter 'h' in them: house, hill, witch, horse, ghost, heart, handkerchief.

CHAPTER FOUR

Page 38 Exercise 1
1. C **2.** B **3.** A **4.** A **5.** B **6.** A
7. C **8.** A **9.** B **10.** A **11.** A

Page 39 Exercise 2
1. A
2. C
3. B
4. C
5. B

Page 40 Exercise 3
1. c
2. f
3. d
4. a
5. b

THE AMERICAN REVOLUTION

Page 41 Exercise 1
1. C
2. B
3. A
4. C
5. B

EXIT TEST

Page 44 Exercise 1
1. f
2. a
3. e
4. b
5. c
6. d

Page 44 Exercise 2
1. T
2. F - Rip does not like working and he is lazy.
3. F - Mrs Van Winkle nags her husband and she is harsh with him.
4. T
5. F - Rip meets a strange old man in the mountains.
6. F - The men are playing a game of ninepins.
7. T
8. T
9. F - After his long sleep everything is changed and Rip does not find his family or his friends.
10. T
11. T

THE LEGEND OF SLEEPY HOLLOW

CHAPTER ONE

Page 50 Exercise 1
1. A
2. B
3. A
4. C
5. B
6. A
7. B

Page 50 Exercise 2
1. c
2. e
3. a
4. b
5. d

CHAPTER TWO

Page 54 Exercise 1
1. c
2. i
3. e
4. a
5. f
6. g
7. b
8. d
9. h
10. j

Page 55 Exercise 2

His arms and legs are very long; His feet are very big. His head is small with big ears. He has green eyes and a long nose.

CHAPTER THREE

Page 59 Exercise 1

1. B
2. C
3. A
4. B
5. C
6. A

Page 60 Exercise 2
d, g, e, c, f, b, a

Page 61 Exercise 3
is, lives, lovely, farm, animals,
prepares, food, admirers, nice,
visits, likes, handsome, Katrina's,
visits, likes, father's, mother's,
arms, legs, looks, does

Page 61 Exercise 4
Who does she love?
a. clothes
b. beautiful
c. rough
d. furious
e. mischievous
f. scare
g. winter
h. broad

NEW YORK IN THE 1800s

Page 63 Exercise 1

```
M  E  A  U  Q  B  O  C  S  B  G  D  T
B  F  P  Z  E  G  G  S  H  U  D  C  G
C  X  P  U  M  R  O  U  A  T  R  O  V
S  H  L  Y  M  Y  N  E  M  T  X  I  P
T  L  E  D  U  P  E  H  P  E  A  R  S
V  I  S  E  I  S  K  L  C  R  B  D  H
U  V  I  P  S  R  D  I  T  M  W  M  Z
Z  P  E  Q  O  E  A  S  N  F  C  I  K
O  N  G  B  D  B  S  O  W  S  K  L  L
A  V  E  G  E  T  A  B  L  E  S  K  N
```

Vegetables/Fruits	Dairy
apples	butter
pears	milk
pumpkins	eggs
vegetables	cheese

CHAPTER FOUR

Page 67 Exercise 1
1. C
2. A
3. B
4. A
5. C
6. A
7. B
8. C
9. B
10. C
11. A
12. C

Page 68 Exercise 2
tonight,
house,
7:15 p.m.,
making,
mother,
fruit cake,
grandmother,
pumpkin pie,
walk,
garden,
friends,
9:30 p.m.,
dance,
sing,
Halloween costume

Page 68 Exercise 3

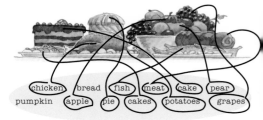

chicken, bread, fish, meat, cake, pear
pumpkin, apple, pie, cakes, potatoes, grapes

CHAPTER FIVE

Page 75 Exercise 1

1. B
2. C
3. A
4. B
5. A
6. B

Page 76 Exercise 2

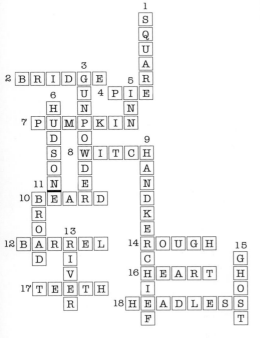

Page 81 Exercise 1

1. C
2. A
3. A
4. B
5. B
6. A
7. B
8. A
9. B
10. A

Page 82 Exercise 2

Open answers.

KATRINA

eighteen years old, beautiful, rich, has a lovely smile, young, wears pretty dresses, has rosy cheeks

ICHABOD

very thin, likes gossiping, always hungry, tall, school teacher, singing teacher, educated, likes reading

BROM

mischievous, young hero, courageous fighter, strong, plays tricks, big, excellent horseback rider

a. Ichabod Crane is a school teacher and a singing teacher.

b. Ichabod likes reading books and eating good food.

c. Katrina is the beautiful daughter of a rich farmer.

d. Both Ichabod and Brom want to marry Katrina.

e. Ichabod and Brom are rivals.

f. Ichabod goes to dinner at the Van Tassels' house.

g. Ichabod meets the Headless Horseman in the forest.

h. The Headless Horseman follows Ichabod.

i. The Headless Horseman throws his head at Ichabod.

j. No one can find Ichabod the next morning.

k. Brom marries Katrina and keeps his secret.

Black Cat English Readers

BLACK CAT ENGLISH CLUB

Membership Application Form

BLACK CAT ENGLISH CLUB is for those who love English reading and seek for better English to share and learn with fun together.

Benefits offered:
- *Membership Card*
- *Member badge, poster, bookmark*
- *Book discount coupon*
- *Black Cat English Reward Scheme*
- *English learning e-forum*
- *Surprise gift and more...*

Simply fill out the application form below and fax it back to 2565 1113.

Join Now! It's FREE exclusively for readers who have purchased *Black Cat English Readers* !

The book(or book set) that you have purchased: _____

English Name: _____ (Surname) _____ (Given Name)

Chinese Name: _____

Address: _____

Tel: _____ Fax: _____

Email: _____

Sex: ❏ Male ❏ Female   (Login password for e-forum will be sent to this email address.)

Education Background: ❏ Primary 1-3   ❏ Primary 4-6   ❏ Junior Secondary Education (F1-3)

❏ Senior Secondary Education (F4-5)   ❏ Matriculation

❏ College   ❏ University or above

Age: ❏ 6 - 9   ❏ 10 - 12   ❏ 13 - 15   ❏ 16 - 18   ❏ 19 - 24   ❏ 25 - 34

❏ 35 - 44   ❏ 45 - 54   ❏ 55 or above

Occupation: ❏ Student   ❏ Teacher   ❏ White Collar   ❏ Blue Collar

❏ Professional   ❏ Manager   ❏ Business Owner   ❏ Housewife

❏ Others (please specify: _____)

As a member, what would you like **BLACK CAT ENGLISH CLUB** to offer:

❏ Member gathering/ party   ❏ English class with native teacher   ❏ English competition

❏ Newsletter   ❏ Online sharing   ❏ Book fair

❏ Book discount   ❏ Others (please specify: _____)

Other suggestions to **BLACK CAT ENGLISH CLUB**:

_____

Please sign here: _____

(Date: _____)